Josiah Parsons Cooke

Scientific culture

.

Josiah Parsons Cooke

Scientific culture

ISBN/EAN: 9783337414191

Printed in Europe, USA, Canada, Australia, Japan

Cover: Foto ©berggeist007 / pixelio.de

More available books at **www.hansebooks.com**

BY

JOSIAH P. COOKE, Jun.

PROFESSOR OF CHEMISTRY AND MINERALOGY IN HARVARD COLLEGE

HENRY S. KING & CO., LONDON

1876

SCIENTIFIC CULTURE.

YOU have come together this morning to begin various elementary courses of instruction in chemistry and mineralogy. As I have been informed, most of you are teachers by profession, and your chief object is to become acquainted with the experimental methods of teaching physical science, and to gain the advantages in your study which the large apparatus of this university is capable of affording.

In all this I hope you will not be disappointed. You, as teachers, know perfectly well that success must depend, first of all, on your own efforts ; but, since the methods of studying Nature are so different from those with which you are familiar in literary studies, I feel that the best service I can render, in this introductory address, is to state, as clearly as I can, the great objects which should be kept in view in the courses on which you are now entering.

By your very attendance on these courses you

have given the strongest evidence of your appreciation of the value of chemical studies as a part of the system of education, and let me say, in the first place, that you have not overvalued their importance. The elementary principles and more conspicuous facts of chemistry are so intimately associated with the experience of every-day life, and find such important applications in the useful arts, that no man at the present day can be regarded as educated who is ignorant of them. Not to know why the fire burns, or how the sulphur-trade affects the industries of the world, will be regarded by the generation of men among whom your pupils will have to win their places in society, as a greater mark of ignorance than a false quantity in Latin prosody or a solecism in grammar.

Moreover, I need not tell you that physical science has become a great power in the world. Indeed, after religion, it is the greatest power of our modern civilisation. Consider how much it has accomplished during the last century toward increasing the comforts and enlarging the intellectual vision of mankind. The railroad, the steamship, the electric telegraph, photography, gas lights, petroleum oils, coal-tar colours, chlorine bleaching, anæsthesia, are a few of its recent material gifts to the world ; and not only has it made one pair of hands to do the work of twenty, but it has so improved and facilitated the old industries, that what were luxuries to the fathers of our republic have become necessities to our generation.

And when, passing from these material fruits, you consider the purely intellectual triumphs of physical science, such as those which have been gained with the telescope, the microscope, and the spectroscope, you cannot wonder at the esteem in which these branches of study are held in this practical age of the world.

Now, these immense results have been gained by the application to the study of Nature of a method which was so admirably described by Lord Bacon in his ' Novum Organon,' and which is now generally called the experimental method. What we observe in Nature is an orderly succession of phenomena. The ancients speculated about these phenomena as well as ourselves, but they contented themselves with speculations, animating Nature with the products of their wild fancies. Their great master, Aristotle, has never been excelled in the art of dialectics ; but his method of logic applied to the external world was of very necessity an utter failure. It is frequently said, in defence of the exclusive study of the records of ancient learning, that they are the products of thinking, loving, and hating men, like ourselves, and it is claimed that the study of science can never rise to the same nobility because it deals only with *lifeless* matter. But this is a mere play on words, a repetition of the error of the old schoolmen.

Physical science is noble because it does deal with thought, and with the very noblest of all thought. Nature at once manifests and conceals an Infinite Presence : Her methods and orderly successions are the manifestations of Omnipotent Will ; Her contrivances and laws the embodiment of Omniscient Thought. The disciples of Aristotle so signally failed simply because they could see in Nature only a reflection of their idle fancies. The followers of Bacon have so gloriously succeeded because they approached Nature as humble students, and, having first learned how to question Her, have been content to be taught and not sought to teach. The ancient logic never relieved a moment of pain, nor lifted an ounce of the burden of human misery. The modern logic has made a very large share of material comfort the common heritage of all civilised men.

In what, then, does this Baconian system consist? Simply in these elements: 1. Careful observation of the conditions under which a given phenomenon occurs; 2. The varying of these conditions by experiments, and observing the effects produced by the variation. We thus find that some of the conditions are merely accidental circumstances, having no necessary connection with the phenomenon, while others are its invariable antecedent. Having now discovered the true relations of the phenomenon we are studying, a happy guess, suggested probably by analogy, furnishes us with a clue to the real causes on which it depends. We next test our guess by further experiments. If our hypothesis is true, this or that must follow; and, if in all points the theory holds, we have discovered the law of which we are in search. If, however, these necessary inferences are not realised, then we must abandon our hypothesis, make another guess, and test that in its turn. Let me illustrate by two well-known examples:

The, of old, universally accepted principle that all living organisms are propagated by seeds or germs (*omnia ex ovo*) has been seriously questioned by a modern school of naturalists. Various observers have maintained that there were conditions under which the lower forms of organic life were developed independently of all such accessories, but other, and equally competent naturalists, who have attempted to investigate the subject, have obtained conflicting results.

Thus it was observed that certain low forms of life were quite constantly developed in beef-juice that had been carefully prepared and hermetically sealed in glass flasks, even after these flasks had been exposed for a long time to the temperature of boiling water. ' Here,' proclaims the new school, ' is unmistakable evidence of spontaneous generation ; for, if past experience is any guide, all germs must have been killed by the boiling water.' 'No,' answer the more cautious naturalists, ' you have not yet proved your point. You have no right to assume that all germs are killed at this temperature.'

The experiments, therefore, were repeated under various conditions and at different temperatures, but with unsatisfactory results, until Pasteur, a distinguished French physicist, devised a very simple mode of testing the question. He reasoned thus : ' If, as is generally believed, the presence of invisible spores in the air is an essential condition of the development of these lower growths, then their production must bear some proportion to the abundance of these spores. Near the habitations of animals and plants where the spores are known to be in abundance, the development would be naturally at a maximum, and we should expect that the growth would diminish in proportion as the microscope indicated that the spores diminished in the atmosphere.'

Accordingly, Pasteur selected a region in the Jura Mountains suitable for his purpose, and repeated the well-known experiment with beef-juice, first at the inn of a town at the foot of the mountains, and then at various elevations up to the bare rocks which covered the top of the ridge, a height of some 8,000 feet. At each point he sealed up beef-juice in a large number of flasks and watched the result. He found that while in the town the animalcules were developed in almost all the flasks, they appeared only in two or three out of a hundred cases where the flasks had been sealed at the top of the mountain, and to a proportionate extent in those sealed at the intermediate elevations. What, now, did these experiments prove? Simply this, that the development of these organic forms was in direct proportion to the number of germs in the air. It did not settle the question of spontaneous generation, but it showed that false conclusions had been deduced from the experiments which had been cited to prove it.

A still more striking illustration of the same method of questioning Nature is to be found in the investigation of Sir Humphry Davy on the composition of water. The voltaic battery which works our telegraphs was invented by Volta in 1800 ; and later, during the same year, it was discovered in London, by Nicholson and Carlisle, that this remarkable instrument had the power of decomposing water. These physicists at once recognised that the chief products of the action of the battery on water were hydrogen and oxygen gases, thus confirming the results of Cavendish, who in 1781 had obtained water by combining these elementary substances ; oxygen having been previously discovered in 1775, and hydrogen at least as early as 1766. It was, however, very soon also observed that there were always formed by the action of the battery on water, besides these aëriform products, an alkali and an acid, the alkali collecting around the negative pole and the acid around the positive pole of the electrical combination. In regard to the nature of this acid and alkali there was the greatest difference of opinion among the early experimenters on this subject. Cruickshank supposed that the acid was nitrous acid, and the alkali ammonia. Desormes, a French chemist, attempted to prove that the acid was muriatic acid ; while Brugnatelli asserted that a new and peculiar acid was formed, which he called the electric acid.

It was in this state of the question that Sir Humphry Davy began his investigation. From the analogies of chemical science, as well as from the previous experiments of Cavendish and Lavoisier, he was persuaded that water consisted solely of oxygen and hydrogen gases, and that the acid and alkali were merely adventitious products. This opinion was undoubtedly well founded ; but, great disciple of Bacon as he was, Davy felt that his opinion was worth nothing unless substantiated by experimental evidence, and accordingly he set himself to work to obtain the required proof.

In Davy's first experiments the two glass tubes which he used to contain the water were connected together by an animal membrane, and he found, on immersing the poles of his battery in their respective tubes, that, besides the now well-known gases, there were really formed muriatic acid in one tube and a fixed alkali in the other. Davy at once, however, suspected that the acid and alkali came from common salt contained in the animal membrane, and he therefore rejected this material and connected the glass tubes by carefully washed cotton fibre : when, on submitting the water as

before to the action of the voltaic current, and con-
tinuing the experiment through a great length of
time, no *muriatic* acid appeared; but he still found
that the water in the one tube was strongly alkaline, and in the other strongly acid, although the
acid was, chiefly at least, nitrous acid. A part of
the acid evidently came from the animal membrane, but not the whole, and the source of the
alkali was as obscure as before.

Davy then made another guess. He knew that
alkali was used in the manufacture of glass; and it
occurred to him that the glass of the tubes, decom-
posed by the electric current, might be the origin
of the alkali in his experiments. He therefore
substituted for the glass tubes cups of agate, which
contains no alkali, and repeated the experiment,
but still the troublesome acid and alkali appeared.
Nevertheless, he said, it is possible that these products may be derived from some impurities exist-
ing in the agate cups, or adhering to them; and
so, in order to make his experiments as refined as
possible, he rejected the agate vessels and procured
two conical cups of pure gold, but on repeating the
experiments the acid and alkali again appeared.

And now let me ask, who is there of us who
would not have concluded at this stage of the in-
quiry that the acid and alkali were essential products of the decomposition of water? But not so
with Davy. He knew perfectly well that all the
circumstances of his experiments had not been
tested, and until this had been done he had no
right to draw such a conclusion. He next turned
to the water he was using. It was distilled water,
which he supposed to be pure, but still, he said, it
is possible that the impurities of the spring-water

may be carried over to a slight extent by the steam in the process of distillation, and may therefore exist in my distilled water to a sufficient amount to have caused the difficulty. Accordingly he evaporated a quart of this water in a silver dish, and obtained seven-tenths of a grain of dry residue. He then added this residue to the small amount of water in the gold cones and again repeated the experiment. The proportion of alkali and acid was sensibly increased.

You think he has found at last the source of the acid and alkali in the impurities of the water. So thought Davy, but he was too faithful a disciple of Bacon to leave this legitimate inference unverified. Accordingly he repeatedly distilled the water from a silver alembic until it left absolutely no residue on evaporation, and then with water which he knew to be pure, and contained in vessels of gold from which he knew it could acquire no taint, he still again repeated the already well-tried experiment. He dipped his test-paper into the vessel connected with the positive pole, and the water was still decidedly acid. He dipped the paper into the vessel connected with the negative pole, and the water was still alkaline.

You might well think that Davy would have been discouraged here. But not in the least. The path to the great truths which Nature hides often leads through a far denser and a more bewildering forest than this ; but then there is not infrequently a *blaze* on the trees which points out the way, although it may require a sharp eye in a clear head to see the marks. And Davy was well enough trained to observe a circumstance which showed that he was now on the right path and heading straight for the goal.

On examining the alkali formed in this last experiment, he found that it was not, as before, a fixed alkali, soda, or potash, but the volatile alkali, ammonia. Evidently the fixed alkali came from the impurities of the water, and when, on repeating the experiment with pure water in agate cups or glass tubes, the same results followed, he felt assured that so much at least had been established. There was still, however, the production of the volatile alkali and of nitrous acid to be accounted for. As these contain only the elements of air and water, Davy thought that possibly they might be formed by the combination of hydrogen at the one pole and of oxygen at the other with the nitrogen of the air, which was necessarily dissolved in the water. In order, therefore, to eliminate the effect of the air, he again repeated the experiment under the receiver of an air-pump from which the atmosphere had been exhausted, but still the acid and alkali appeared in the two cups.

Davy, however, was not discouraged by this, for the *blazes* on the trees were becoming more numerous, and he now felt sure that he was fast approaching the end. He observed that the quantity of acid and alkali had been greatly diminished by exhausting the air, and this was all that could be expected, for, as Davy knew perfectly well, the best air-pumps do not remove all the air. He therefore for the last experiment not only exhausted the air, but replaced it with pure hydrogen, and then exhausted the hydrogen and refilled the receiver with the same gas several times in succession, until he was perfectly sure that the last traces of air had been, as it were, washed out. In this atmosphere of pure hydrogen he allowed the battery to act on the water, and not until the end of twenty-four hours did he disconnect the apparatus. He then dips his test-paper into the water connected with the positive pole, and there is no trace of acid; he dips it into the water at the negative pole, and there is no alkali; and you may judge with what satisfaction he withdraws those slips of test-paper, whose unaltered surfaces showed that he had been guided at last to the truth, and that his perseverance had been rewarded.

The fame of Sir Humphry Davy rests on his discovery of the metals of the alkalies and earths which first revealed the wonderful truth that the crust of our globe consists of metallic cinders; but none of these brilliant results show so great scientific merit or such eminent power of investigating Nature as the experiments which I have just detailed. I have not, however, described them here for the purpose of glorifying that renowned man. His honoured memory needs no such office at my hands. My only object was to show you what is meant by the Baconian method of science, and to give some idea of the nature of that modern logic which within the last fifty years has produced more wonderful transformations in human society than the author of Aladdin ever imagined in his wildest dreams. In this short address I can of course give you but a very dim and imperfect idea of what I have called the Baconian system of experimental reasoning. Indeed, you cannot form any clear conception of it, until in some humble way you have attempted to use the method, each one for himself, and you have come here in order that you may acquire such experience.

My object, however, will be gained if these illustrations serve to give emphasis to the following statements, which I feel I ought to make at the opening of these courses of instruction—statements which have an especial appropriateness in this place ; since I am addressing teachers who are in a position to exert an important influence on the system of education in this country.

In the first place, then, I must declare my conviction that no educated man can expect to realise his best possibilities of usefulness without a practical knowledge of the methods of experimental science. If he is to be a physician, his whole success will depend on the skill with which he can use these great tools of modern civilisation. If he is to be a lawyer, his advancement will in no small measure be determined by the acuteness with which he can criticise the manner in which the same tools have been used by his own or his opponent's clients. If he is to be a clergyman, he must take sides in the great conflict between theology and science, which is now raging in the world, and, unless he wishes to play the part of the doughty knight Don Quixote, and think he is winning great victories by knocking down the imaginary adversaries which his ignorance has set up, he must try the steel of his adversary's blade.

Let me be fully understood. It is not to be expected or desired that many of our students should become professional men of science. The places of employment for scientific men are but few, and more in the future than in the past they will naturally be secured by those whom Nature has endowed with special aptitudes or tastes—usually the signs of aptitudes—to investigate her laws. That our country will always offer an honourable career to her men of genius, we have every reason to expect, and these born students of Nature will usually follow the plain indications of Providence without encouragement or direction from us.

It is different, however, with the great body of earnest students who are conscious of no special aptitudes, but who are desirous of doing the best thing to fit themselves for usefulness in the world ; and I feel that any system of education is radically defective which does not comprise a sufficient training in the methods of experimental science to make the mass of our educated men familiar with this tool of modern civilisation : so that when, hereafter, new conquests over matter are announced, and great discoveries are proclaimed, they may be able not only to understand but also to criticise the methods by which the assumed results have been reached, and thus be in a position to distinguish between the true and the false. Whether we will or not, we must live under the direction of this great power of modern society, and the only question is, whether we will be its ignorant slave or its intelligent servant.

In the second place, it seems fitting that I should state to you what I regard as the true aims to be kept in view in a course of scientific study, and to give my reasons for the methods we have adopted in arranging the courses you are about to begin.

In our day there has arisen a warm discussion as to the relative claims of two kinds of culture, and attempts are made to create an antagonism between them. But all culture is the same in spirit. Its object is to awaken and strengthen the

B

powers of the mind ; for these, like the muscles of
the body, are developed and rendered strong and
active only by exercise ; while on the other hand
they may become atrophied from mere want of
use.   Science culture differs in its methods from
the old classical culture, but it has the same spirit
and the same object.   You must not, therefore, expect me to advocate the former at the expense of
the latter ; for, although I have laboured assiduously
during a quarter of a century to establish the
methods of science-teaching which have now become general, I am far from believing that they
are the only true modes of obtaining a liberal
education.   So far from this, if it were necessary
to choose one of two systems, I should favour the
classical ; and why ?

Language is the medium of thought, and cannot be separated from it.   He who would think
well must have a good command of language, and
he who has the best command of language I am
almost tempted to say will think the best.   For this
reason a certain amount of critical study of language
is essential for every educated man, and such study
is not likely to be gained except through the great
ancient languages ; the advocates of classical
scholarship frequently say, cannot be gained.   I
am not ready to accept this dictum ; but I most
willingly concede that in the present state of our
schools it is not likely to be gained.   I never had
any taste myself for classical studies ; but I know
that I owe to the study a great part of the mental
culture which has enabled me to do the work that
has fallen to my share in life.

But while I concede to all this, I do not believe,
on the other hand, that the classical is the only

effective method of culture; you evidently do not think so, for you would not be here if you did. But; in abandoning the old tried method, which is known to be good, for the new, you must be careful that you gain the advantages which the new offers; and you will not gain the new culture you seek unless you study science in the right way. In the classical departments the methods are so well established, and have been so long tested by experience, that there can hardly be a wrong way. But in science there is not only a wrong way, but this wrong way is so easy and alluring, that you will most certainly stray into it unless you strive earnestly to keep out of it. Hence I am most anxious to point out to you the right way, and do what I can to keep you in it; and you will find that our courses and methods have been devised with this object.

When advocating in our mother University of Cambridge, in old England, the claims of scientific culture, I was pushed with an argument which had very great weight with the eminent English scholars present, and which you will be surprised to learn was regarded as fatal to the success of the science *triposes* then under debate. The argument was, that the experimental sciences could not be made the subjects of competitive examinations. Some may smile at such an objection; but, as viewed from the English standpoint, there was really a great deal in it, and the argument brought out the radical difference between scientific and classical culture.

The old method of culture may be said to have culminated in the competitive examinations of the English universities. We have no such examinations here. Success depends not simply on knowing your subject thoroughly, but on having it at your fingers' ends, and those fingers so agile that they can accomplish not only a prodigious amount of work in a short time, but can do this work with absolute accuracy. For the only approach we make to an experience of this kind, we must look to our athletic contests. It may of course be doubted whether the ability, once in a man's life, to perform such mental feats, is worth what it costs. Still it implies a very high degree of mental culture, and it is perfectly certain that the experimental sciences give no field for that sort of mental prize-fights. It is easy to prepare written examinations which will show whether the students have been faithful to their work, but they cannot be adapted to such competitions as I have described without abandoning the true object of science culture. The ability of the scientific student can only be shown by long-continued work at the laboratory table, and by his success in investigating the problems which Nature presents.

We have here struck the true key-note of the scientific method. The great object of all our study should be to study Nature, and all our methods should be directed to this one object. This aim alone will ennoble our scholarship as students, and will give dignity to our scientific calling as men of science. It is this high aim, moreover, which vindicates the worth of the mode of culture we have chosen. What is it that ennobles literary culture, but the great minds which, through this culture, have honoured the nations to which they belong ?

The culture we have chosen is capable of even greater things ; not because science is nobler than art, for both are equally noble ;—it is the thought, the conception, which ennobles, and I care not whether it be attained through one kind of exercise of the mental faculties or another ;—but we are capable of grander and nobler thoughts than Plato, Cicero, Shakspeare, or Newton, because we live in a later period of the world's history ; when, through science, the world has become richer in great ideas. It is, I repeat, the great thought which ennobles, and it ennobles because it raises to a higher plane that which is immortal in our manhood.

If I have made my meaning clear, and if you sympathise with my feelings, you will understand why I regard culture as so important to the individual and to the nation. The works of Shakspeare and of Bacon are of more value to England to-day than are the memories of Blenheim or Trafalgar ; and those great minds will still be living powers in the world when Marlborough and Nelson are only remembered as historical names.

I therefore believe that it is the first duty of a country to foster the highest culture, and that it should be the aim of every scholar to promote this culture both by his own efforts and his active influence. A nation can become really great in no other way. We live in a country of great possibilities; and the danger is that, as with many men I have known in college, of great potential abilities, the greatness will end where it begins. The scholars of the country should have but one voice in this matter, and urge upon the Government and upon individuals the duty of encouraging and supporting mental culture for its own sake.

The time has passed when we can afford to limit the work of our higher institutions of learning to teaching knowledge already acquired. Henceforth the investigation of unsolved problems, and the discovery of new truth, should be one of the main objects at our universities, and no cost should be grudged, which is required to maintain at them the most active minds, in every branch of knowledge, which the country can be stimulated to produce.

I could urge this on the self-interest of the nation as an obvious dictate of political economy. I could say, and say truly, that the culture of science will help us to develop those latent resources of which we are so proud; will enable us to grow two blades of grass where one grew before; to extract a larger per cent. of metal from our ores; to economise our coal, and in general to direct our waiting energies so that they may produce a more abundant pecuniary reward. I could tell of Galvani studying for twenty long years, to no apparent purpose, the twitching of frogs' hind-legs, and thus sowing the seed from which has sprung the greatest invention of modern times. Or, if our Yankee impatience would be unwilling to wait half a century for the fruit to ripen, I could point to the purely theoretical investigations of organic chemistry, which in less than five years have revolutionised one of the great industries of Europe, and liberated thousands of acres for a more beneficent agriculture. This is all true, and may be urged properly if higher considerations will not prevail. It is an argument I have used in other places, but I will not use it here; although I gladly acknowledge the Providence which brings at last even material fruits to reward conscientious labour for the advancement of knowledge and the intellectual elevation of mankind. I would rather point to that far greater multitude who have worked in faith for the love of knowledge, and who have ennobled themselves and ennobled their nation, not because they have added to its material prosperity, but because they have made themselves and made their fellows more noble men.

I come back now again to the moral of all this, to urge upon you, as the noblest patriotism and the most enlightened self-interest, the duty of striving for yourselves and encouraging in others the highest culture in the studies you have chosen, and this culture with one end in view—to advance knowledge. I am far, of course, from advising you to grapple immaturely with unsolved problems, or, when you have gained the knowledge with which you can dare to venture from the beaten track, to undertake work beyond your power. Many a young scientific man has suffered the fate of Icarus in attempting to soar too high. Moreover, I am far from expecting that all or many of you will ever have the opportunity of going beyond the well-explored fields of knowledge ; but you can all have the aim, and that aim will make your work more worthy and more profitable to yourselves. Every American boy cannot be President of the United States, but if, as our English cousins allege, he believes that he can be, the very belief makes him an abler man.

We have dwelt long enough on these generalities, and it is time to come down to commonplaces, and to inquire what are the essential conditions of this scientific culture which shall fit us to investigate Nature ; and the first thought that occurs to me in this connection may be expressed thus : Science presents to us two aspects, which I may call its objective and its subjective aspect. Objectively it is a body of facts, which we have to observe, and subjectively it is a body of truths, conclusions, or inferences, deduced from these facts ; and the two sides of the subject should always be kept in view.

I propose next to say a few words in regard to each of these two aspects of our study, and in regard to the best means of training our faculties so as to work successfully in each sphere. First, then, success in the observation of phenomena implies three qualities at least, namely, quickness and sharpness of perception, accuracy in details, and truthfulness ; and on its power to cultivate these qualities a large part of the value of science, as a means of education, depends.

To begin with the cultivation of our perceptions. We are all gifted with senses, but how few of us use them to the best advantage ! 'We have eyes and see not ;' for, although the light paints the picture on the retina, our dull perceptions give no attention to the details, and we retain only a confused impression of what has passed before our eyes. 'But how,' you may ask, 'are we to cultivate this sharpness of perception ?' I answer, only by making a conscious effort to fix our attention on the objects we study, until the habit becomes a second nature. I have often noticed with surprise, the power which uneducated miners frequently possess, of recognising many minerals at sight. This they have acquired by long experience and close familiarity with such objects, and such power of observation is with them so purely a habit that they are frequently unable to

state clearly the grounds on which their conclusions
are based. They recognise the minerals by what
in common language is called their *looks*, and they
notice delicate differences in the *looks* to which
most men are blind. It is, however, the business
of the scientific mineralogist to analyse these *looks*,
and to point out in what the differences consist; so
that by fixing his attention on these points the
student may gain, by a few hours' study, the power
which the miner acquires only after long experience.

The chief difficulty, however, which we find in
teaching mineralogy is, that the students do not
readily see the differences when they are pointed
out, or, if they see them, do not remember them
with sufficient precision to render their subsequent
observations conclusive and precise. This either
arises from a failure to cultivate the powers of observation in childhood, or the subsequent blunting
of them by disuse. Ladies will scout the idea
that a brooch of cut-glass is as ornamental as
one of diamond, and yet I venture to assert
that there is not one person in fifty, at least of
those who have not made a study of the subject, who can tell the difference between the
two. The external appearance depends simply
on what we call lustre. The lustre of glass is vit-
reous, that of the diamond adamantine, and I know,
of no other distinction which it is more difficult for
students to recognise than this.,. Those of you who
study mineralogy will experience this difficulty, and
it can be overcome only by giving careful attention
to the subject. The teacher can do nothing more
than put in your hands the specimens which illustrate the point, and you must study these specimens

until you see the difference. It is a question of
sight, not of understanding, and all the optical
theories of the cause of the lustre will not help you
in the least toward seeing the difference between
diamond and glass, or anglesite and heavy spar.

Another illustration of the same fact is the con-
stant failure of students to distinguish by the eye
alone between the two minerals called copper-glance
and gray copper. There is a difference of colour
and lustre which, although usually well marked, it
requires an educated eye to distinguish.

Mineralogy undoubtedly demands a more careful cultivation of the perceptions than the other
branches of chemistry; but still you will find
abundant practice for close observation in them all.
I have often known students to reach erroneous
results in qualitative analysis by mistaking a white
precipitate in a coloured liquid for a coloured precipitate ; or by not attending to similar broad dis-
tinctions which would have been obvious to any
careful observer ; and so in quantitative analysis,
mere delicacy of touch or handling is a great
element of success.

But I must pass on to speak of the importance
in the study of Nature of accuracy in detail, which
is the second condition of successful observation of
which I spoke. We must cultivate not only accuracy in observing details, but also accuracy in following details which have been laid down by others
for our guidance. In science we cannot draw correct conclusions from our premises unless we are
sure that we have all the facts, and what seemed at
first an unimportant detail often proves to be the
determining condition of the result ; and, again,
if we are told that under certain conditions a cer-

tain sign is the proof of the presence of a certain substance, we have no right to assume that the sign is of any value unless the conditions are fulfilled. A black precipitate, for example, obtained under certain conditions, is a proof of the presence of nickel, but we cannot assert that we have found nickel unless we have followed out those details in every particular.

Of course, we must avoid empiricism as far as we can. We must seek to learn the reasons of the details, and such knowledge will not only render our works intelligent, but will also frequently enable us to judge how far the details are essential, and to what extent our processes may be varied with safety. We must also avoid trifling, and above all 'the straining at a gnat and swallowing a camel,' as is the habit with triflers. Large knowledge and good judgment will avoid all such errors ; but, if we must choose between fussiness and carelessness, the first is the least evil. Slovenly work means slovenly results, and habits of carefulness, neatness, and order, produce as excellent fruits in the laboratory as in the home.

Last in order but first in importance of the conditions of successful observation, mentioned above, stands truthfulness. Here you may think I am approaching a delicate subject, of which even to speak might seem to cast a reproach. But not so at all. I am not speaking here of conscious deception, for I assume that no one who aspires to be a student of Nature can be guilty of that. But I am speaking of a quality whose absence is not necessarily a mark of sinfulness, but whose possession, in a high degree, is a characteristic of the greatest scientific talent. As every lawyer knows,

he is a rare man whose testimony is not coloured
by his interests, and a very large amount of self-
deception is compatible with conscious honesty of
purpose.

So among scientific students the power to keep
the mind unbiassed and not to colour our observations in the least degree, is one of the rarest, as it is
one of the noblest of qualities.   It is a quality we
must strive after with all our might, and we shall
not attain it unless we strive.   Remember, our observations are our data, and, unless accurate, every
thing deduced from them must have the taint of
our deception.   We cannot deceive Nature, however much we may deceive ourselves ; and there is
many a student who would cut off his right hand
rather than be guilty of a conscious untruth, who
is yet constantly untruthful to himself.   Every
year students of mineralogy present to me written
descriptions of mineral specimens which particula-
rise, as observed, characters that do not appear on
the specimens given them to determine, although
they may be the correct characters of some other
mineral.

There is usually no want of honesty in this, but,
deceived by some accident, the student has made a
wrong guess, and then imagined that he saw on the
specimen those characters which he knew from the
descriptions ought to appear on the assumed mineral.
So, also, it not unfrequently happens that a student
in qualitative analysis, who has obtained some hints
in regard to the composition of his solution, will
torture his observations until they seem to him to
confirm his erroneous inferences ; and again the
student in quantitative analysis, who finds out the
exact weight he ought to obtain, is often insensibly

influenced by this knowledge—in the washing and ignition of his precipitate, or in some other way—and thus obtains results whose only apparent fault may be a too close agreement with theory, but which, nevertheless, are not accurate because not true. It is evident how fatal such faults as these must be to the investigation of truth, and they are equally destructive of all scientific scholarship. Their effect on the student is so marked, that although he may deceive himself, he will rarely deceive his teacher. That he should lose confidence in his own results is, to the teacher, one of the most marked indications of such false methods of study, but the student usually refers his want of success to any cause but the real one—his own untruthfulness. He will complain of the teacher, or of the methods of instruction, and may even persuade himself that all scientific results are as uncertain as his own. As I have said, mere ordinary truthfulness, which spurns any conscious deception, will not save us from falling into such faults. Our scientific study demands a much higher order of truthfulness than this. We should so love the truth above all price, as to strive for it with single-hearted and unswerving purpose. We must be constantly on our guard to avoid any circumstance which would tend to bias our minds or warp our judgments, and we must make the at-tainment of the truth our sole motive guide and end.

It remains for me, before closing this address, to say a few words on what I have called the subjective aspect of scientific study. Science offers us not only a mass of phenomena to be observed, but also a body of truths which have been deduced from these observations ; and, without the power of drawing correct inferences from the data acquired, exact observations would be of little value. I have already described the inductive method of reasoning, and illustrated it by two noteworthy examples, and, in a humbler measure, we must apply the same method in our daily work in the laboratory. We must learn how to vary our experiments so as to eliminate the accidental circumstances, and make evident the essential conditions of the phenomena we are studying. Such power can only be acquired by practice, and a somewhat long experience in active teaching has convinced me that there is no better means of training this logical faculty than the study of qualitative chemical analysis in which many of you are to engage.

The results of the processes of qualitative analysis are perfectly definite and trustworthy; but they are only reached by following out the indications of experiments which are frequently obscure, and even apparently contradictory; reconciling by new experiments the seeming discrepancies, and, at last, having eliminated all other possible causes of the phenomena observed, discovering the true nature of the substances under examination.

The study of mineralogy affords an almost equally good practice, although in a somewhat different form. By comparing carefully many specimens of the same mineral, you learn to distinguish the accidental from the essential characters, and on this distinction you must base your inferences in regard to the nature of the specimens you may be called upon to determine. A single remark occurs to me which may aid you in cultivating this scientific logic.

Do not attempt to reason on insufficient data. Multiply your observations or experiments, and, when your premises are ample, the conclusion will generally take care of itself. Are you in doubt in regard to a mineral specimen? Repeat your observations again and again, multiply them with the aid of the blow-pipe or goniometer, compare the specimen with known specimens which it resembles, until either your doubts are removed, or you are satisfied that you are unequal to the task; and remember that, in many cases, the last is the only honest conclusion.

Are you in doubt in regard to the reactions of the substance you are analysing, whether they are really those of a metal you suspect to be present? Do not rest in such a frame of mind, and, above all, do not try to remove the doubt by comparing your experience with that of your neighbour: but multiply your own experiments; procure some compound of the metal, and compare its reactions with those you have observed, until you reach either a positive or a negative result.

Remember that the way to remove your doubts is to widen your own knowledge, and not to depend on the knowledge of others. When your knowledge of the facts is ample, your inferences will be satisfactory, and then an unexplained phenomenon is the guide to a new discovery. Do not be discouraged if you have to labour long in the dark before the day begins to dawn. It will at last dawn to you, as it has dawned to others before, and, when the morning breaks, you will be satisfied with the result of your labour.

Moreover, I feel confident that such experience will very greatly tend to increase your appreciation of the value of scientific studies in training the reasoning faculties of the mind. This, as every one must admit, is the best test of their utility in a scheme of education, and it is precisely here that I claim for them the very highest place. It has generally been admitted that mathematical studies are peculiarly well adapted to train the logical faculties, but still many persons have maintained that, since the mathematics deal wholly with absolute certainties, an exclusive devotion to this class of subjects unfits the mind for weighing the probable evidence by which men are chiefly guided in the affairs of life.

But, without attempting to discuss this question, on which much might be said on both sides, it is certain that no such objection can be urged against the study of the physical sciences if conducted in the manner I have attempted to describe. These subjects present to the consideration of the student every degree of probable evidence, accustoming him to weigh all the evidence for or against a given conclusion, and to reject or to provisionally accept only on the balance of probabilities. Moreover, in practical science, the student is taught to follow out a chain of probable evidence with care and caution, to eliminate all accidental phenomena, and supply, by experiment or observation, the missing links, until he reaches the final conclusion —an intellectual process which, though based wholly on probable evidence, may have all the force and certainty of a mathematical demonstration.

Indeed, that highly-valued scientific acumen and skill which enables the student to brush away the accidental circumstances by which the laws of Nature are always concealed until the truth stands out in bold relief, is but a higher phase of the same talent, which marks professional skill in all the higher walks of life. The physician who looks

through the external symptoms of his patient to
the real disease which lurks beneath ; the lawyer,
who disentangles a mass of conflicting testimony, and follows out the truth successfully
to the end ; the statesman, who sees beneath the
froth of political life the great fundamental principles which will inevitably rule the conduct of the
State, and thus foresees and provides for the coming
change ; the general, who discovers amid the con-
fusion of the battle-field the weak point of his
enemy's front : the merchant, even, who can inter-
pret the signs of the unsettled market—employ
the same faculty, and frequently in not a much
lower degree, that discovered the law of gravita-
tion, and which, since the days of Newton, has
worked so successfully to unveil the mysteries of
the material creation.

Moreover, I hope that you will come to value
scientific studies, not simply because they cultivate
the perceptive and reasoning faculties, but also because they fill the mind with lofty ideals, elevated
conceptions, and noble thoughts. Indeed, I claim
that there is no better school in which to train the
æsthetical faculties of the mind, the tastes, and the
imagination, than the study of natural science.
The beauty of Nature is infinite, and the more we
study her works the more her loveliness unfolds.
The upheaved mountain, with its mantle of eternal
snow ; the majestic cataract, with its whirl and
roar of waters ; the sunset cloud, with its blending
of gorgeous hues, lose nothing of their beauty for
him who knows the mystery they conceal. On the
contrary, they become, one and all, irradiated by
the Infinite Presence which shines through them,
and fill the mind with grander conceptions and

C

nobler ideas than your uneducated child of Nature could ever attain.

Remember that I am not recommending an exclusive devotion to the natural sciences. I am only claiming for them their proper place in the scheme of education, and I do not, of course, deny the unquestionable value of both the ancient and the modern classics in cultivating a pure and elevated taste. But I do say that the poet-laureate of England has drawn a deeper inspiration from Nature interpreted by science than any of his predecessors of the classical school; and I do also affirm that the pre-Raphaelite school of painting, with all its grotesque mimicry of Nature, embodies a truer and purer ideal than that of any Roman fable or Grecian dream.

And what shall we say of the imagination? Where can you find a wider field for its exercise than that opened by the discoveries of modern science? And as the mind wanders over the vast expanse, crossing boundless spaces, dwelling in illimitable time, witnessing the displays of immeasurable power, and studying the adaptations of omniscient skill, it lives in a realm of beauty, of wonder, and of awe, such as no artist has ever attained to in word, in sound, in colour, or in form. And if such a life does not lead man to feel his own dependence, to yearn toward the Infinite Father, and to rest on the bosom of Infinite Love, it is simply because it is not the noble in intellect, not the great in talent, not the profound in knowledge, not the rich in experience, not the lofty in aspiration, not the gifted in imagery, but solely the pure in heart, who see God.

Such, then, is a very imperfect presentation of

what I believe to be the value of scientific studies as a means of education. In what I have stated I have implied that, for these studies to be of any real value, the end must be constantly kept in view, and everything made subservient to the one great object.

To study the natural sciences merely as a collection of interesting facts which it is well for every educated man to know, seldom serves a useful purpose. The young mind becomes wearied with details, and soon forgets what it has never more than half acquired. The lessons become an exercise of the memory and of nothing more; and if, as is too frequently the case, an attempt is made to cram the half-formed mind in a single school-year with an epitome of half the natural sciences—natural philosophy, astronomy, and chemistry, physiology, zoology, botany, and mineralogy, following each other in rapid succession—these studies become a great evil, an actual nuisance, which I should be the first to vote to abate. The tone of mind is not only not improved, but seriously impaired, and the best product is a superficial, smattering smartness, which is the crying evil not only of our schools, but also of our country.

In order that the sciences should be of value in our educational system, they must be taught more from things than from books, and *never* from books without the things. They must be taught, also, by real living teachers, who are themselves interested in what they teach, are interested also in their pupils, and understand how to direct them aright. Above all, the teachers must see to it that their pupils study with the understanding and not solely with the memory, not permitting a single lesson to be recited which is not thoroughly understood, taking the greatest care not to load the memory with any useless lumber, and eschewing *merely* memorised rules as they would deadly poison. The great difficulty against which the teachers of natural science have to contend in the colleges are the wretched tread-mill habits the students bring with them from the schools. Allow our students to memorise their lessons, and they will appear respectably well, but you might as easily remove a mountain as to make many of them think. They will solve an involved equation of algebra readily enough so long as they can do it by turning their mental crank, when they will break down on the simplest practical problem of arithmetic which requires of them only thought enough to decide whether they shall multiply or divide.

Many a boy of good capabilities has been irretrievably ruined, as a scholar, by being compelled to learn the Latin grammar by rote at an age when he was incapable of understanding it; and I fear that schools may still be found where young minds are tortured by this stupefying exercise. Those of us who have faith in the educational value of scientific studies are most anxious that the students who resort to our colleges should be as well fitted in the physical sciences as in the classics, for otherwise the best results of scientific culture cannot be expected. As it is, our students come to the university, not only with no preparation in physical science, but with their perceptive and reasoning faculties so undeveloped that the acquisition of the elementary principles of science is burdensome and distasteful: and good scholars, who are ambitious of distinction, can more readily win their laurels on the old familiar track than on an untried course of which they know nothing, and for which they must begin their training anew.

We have improved our system of instruction in the college as fast as we could obtain the means, but we are persuaded that the best results cannot be reached without the cooperation of the schools. We feel, therefore, that it is incumbent upon us, in the first place, to do everything in our power to prove to the teachers of this country how great is the educational value of the physical sciences, when properly taught ; and, secondly, to aid them in acquiring the best methods of teaching these subjects. It is with such aims that our summer courses have been instituted, and your presence here in such numbers is the best evidence that they have met a real want of the community. We welcome you to the university and to such advantages as it can afford, and we shall do all in our power to render your brief residence here fruitful both in experience and in knowledge ; hoping also that the university may become to you, as she has to so many others, a bright light shining calmly over the troubled sea of active life, ever suggesting lofty thoughts, encouraging noble endeavours, and inciting all her children to work together toward those great ends, the advancement of knowledge and the education of mankind.

LONDON : PRINTED BY
SPOTTISWOODE AND CO., NEW-STREET SQUARE
AND PARLIAMENT STREET

LIST OF SCIENTIFIC WORKS

PUBLISHED BY

HENRY S. KING & CO.

THE INTERNATIONAL SCIENTIFIC SERIES.

I

The FORMS of WATER in CLOUDS and RIVERS, ICE and GLACIERS. By J. TYNDALL, LL.D., F.R.S. With 26 Illustrations. Fifth Edition. Crown 8vo. 5s.

'A fascinating book dealing largely with the phenomena of glaciers and snow, with which, more than any living man, Professor Tyndall has made us familiar.' *British Quarterly Review.*

II

PHYSICS and POLITICS; or, Thoughts on the Application of the Principles of 'Natural Selection' and 'Inheritance' to Political Society. By WALTER BAGEHOT. Third Edition. Crown 8vo. 4s.

'We can recommend the book as well deserving to be read.'—*Saturday Review.*

'A work of really original and interesting speculation.'—*Guardian.*

III

FOODS. By EDWARD SMITH, M.D., LL.B., F.R.S. Third Edition, profusely Illustrated. 5s.

'Few men have so fully and carefully studied the action of foods as Dr. Edward Smith, and the number of experiments which he has made is quite astonishing. The book contains a very large amount of useful information in a small space.'—*Academy.*

IV

MIND and BODY: the Theories of their Relation. By ALEXANDER BAIN, LL.D. Third Edition. Crown 8vo. 4s.

'The work seeks to complete the doctrine of the relation between consciousness and bodily organism. It proposes to show how completely all the familiar processes of sensation, thought, and emotion flow upon the surface of nervous currents, which sustain and fashion their ever-varying shapes.'—*Saturday Review.*

THE INTERNATIONAL SCIENTIFIC SERIES—(continued).

v

STUDY of SOCIOLOGY. By Herbert Spencer. Fourth Edition. Crown 8vo. 5s.

'There is not a line of Mr. Spencer's which is not worth reading.'—*Examiner.*

' It contains a great amount of interesting and suggestive matter.'
Saturday Review.

vi

On the CONSERVATION of ENERGY. By Professor Balfour Stewart. Third Edition. With 14 Engravings. 5s.

'Thorough and simple. . . . A boon to science and the world at large.'
Saturday Review.

' A lucid and extremely simple exposition.'—*Edinburgh Medical Journal.*

vii

ANIMAL LOCOMOTION; or, Walking, Swimming, and Flying. By J. B. Pettigrew, M.D., F.R.S. Second Edition. With 119 Illustrations. 5s.

'A comprehensive *résumé* of the present advanced state of our knowledge of animal locomotion.'—*Standard.*

' We have great pleasure in recommending it as in every way worthy of perusal. The reader will find it replete with bold, original, and interesting matter.'—*Lancet.*

viii

RESPONSIBILITY in MENTAL DISEASE. By Dr. Henry Maudsley. Second Edition. 5s.

'The volume is altogether one of the best of the International Scientific Series which has yet appeared.'—*Academy.*

' There is much wisdom in this valuable book.'—*Spectator.*

ix

The NEW CHEMISTRY. By Professor Josiah P. Cooke, of the Harvard University. Second Edition. With 31 Illustrations. 5s.

'The great ideas of modern chemistry are presented with singular clearness and with very varied illustration.'—*Lancet.*

x

The SCIENCE of LAW. By Sheldon Amos. Second Edition. Crown 8vo. 5s.

' The object of this work is to bring before the student the results of the investigations and labours of Bentham, Austin, and Sir Henry Maine, so far as they contribute to the advancement of legal knowledge. . . . This we can say, that it does "open out to novices an unsuspected region of interest," and that we hope with Mr. Amos it will "whet their curiosity and stimulate them to further research."'
Law Times.

THE INTERNATIONAL SCIENTIFIC SERIES—(continued).

XI

ANIMAL MECHANISM. By E. J. MAREY. A Treatise on Terrestrial and Aerial Locomotion. Second Edition. With 117 Illustrations. 5*s.*

'A thorough exposition of intricate problems of mechanical physiology, which have been worked out with a degree of ability rarely to be found in a single author.'
Nature.

XII

The DOCTRINE of DESCENT and DARWINISM. By Professor OSCAR SCHMIDT, Strasburg University. Second Edition. 26 Illustrations. 5*s.*

'The book is one of very conspicious ability.'—*Scotsman.*

XIII

HISTORY of the CONFLICT between RELIGION and SCIENCE. By JOHN WILLIAM DRAPER, M.D., LL.D., Professor in the University of New York, Author of 'A Treatise on Human Physiology.' Fifth Edition. 5*s.*

'Certainly it is a long time since a book with such an important bearing on Society, and so thoroughly scientific, was placed in the hands of the British public.'
Popular Science Review.

XIV

FUNGI ; their Nature, Influences, Uses, &c. By M. C. COOKE, M.A., LL.D. Edited by the Rev. M. J. BERKELEY, M.A., F.L.S. Second Edition. With Illustrations. Crown 8vo. 5*s.*

'An excellent and readable introduction to the study of fungi.'—*Field.*

'An admirable handbook for the student.'—*Graphic.*

'Mr. Cooke, whose numerous works on botany have been gratefully received by a large number of students, has done well to put a mass of valuable facts about Fungi into portable shape. His book is in all respects useful and satisfactory.'
Pall Mall Gazette.

XV

The CHEMICAL EFFECTS of LIGHT and PHOTO- GRAPHY, in their application to Art, Science, and Industry. By Professor VOGEL, Polytechnic Academy of Berlin. Second and thoroughly revised Edition. With 100 Illustrations, including some beautiful Specimens of Photography. 5*s.*

THE INTERNATIONAL SCIENTIFIC SERIES—(continued).

XVI

The LIFE and GROWTH of LANGUAGE. By WILLIAM DWIGHT WHITNEY, Professor of Sanskrit and Comparative Philology in Yale College, New Haven. Second Edition. Crown 8vo. 5*s. Copyright Edition.*

' Sure to command the attention and interest of every student of language. . . . The author begins by defining language as "the means of expression of human thought," then describes the way in which each individual has to learn his mother tongue. This leads him on to describe the conservative and alterative forces which change the outer form and inner content of words and so bring about the growth of language.'—*Academy.*

'The keen common sense of Professor Whitney clears the jungle of mystification, and by giving an open view of the ground of discussion at once indicates the way by which to reach the truth. . . . A masterly review of the various methods by which language is enriched and embellished.'—*Examiner.*

XVII

MONEY and the MECHANISM of EXCHANGE. By Professor W. STANLEY JEVONS. Second Edition. Crown 8vo. 5*s.*

' A very readable and interesting—we may almost say—amusing treatise on what is usually a very dismal and bewildering subject—Money.'—*Saturday Review.*

'A collection of facts relative to the past and present monetary systems of the world. . . . . will prove a useful addition to the literature bearing on financial subjects.'—*Athenæum.*

XVIII

The NATURE of LIGHT. With a General Account of Physical Optics. By Dr. EUGENE LOMMEL, Professor of Physics in the University of Erlangen. With 188 Illustrations and a Spectra in Chromolithography. 5*s.*

XIX

ANIMAL PARASITES and MESSMATES. By Monsieur VAN BENEDEN, Professor of the University of Louvain, Correspondent of the Institute of France. With 83 Illustrations. Crown 8vo. 5*s.*

FORTHCOMING VOLUMES.

The FIVE SENSES of MAN. By Professor BERNSTEIN, of the University of Halle. Crown 8vo.

FERMENTATIONS. By Professor SCHUTZENBERGER, Director of the Chemical Laboratory at the Sorbonne. Crown 8vo.

FIELD and FOREST RAMBLES of a NATURALIST
in New Brunswick. By A. L. ADAMS, M.A. With Notes and Observations on the Natural History of Eastern Canada. Illustrated. 8vo. cloth, 14*s.*

'Both sportsmen and naturalists will find this work replete with anecdote and carefully-recorded observation, which will entertain them.'—*Nature.*

'Will be found interesting by those who take a pleasure either in sport or natural history.'—*Athenæum.*

The SCIENTIFIC SOCIETIES of LONDON. By
BERNHARD H. BECKER. 1 vol. crown 8vo. 5*s.*

'This is an interesting and useful volume. Mr. Becker's book is all that can be desired; his descriptions are carefully and well written, and his criticisms for the most part just.'—*Spectator.*

'These sketches are popularly written, and are calculated to excite in many minds a desire for further information.—*Daily News.*

STUDIES of BLAST FURNACE PHENOMENA.
By M. L. GRUNER. Translated by L. D. B. GORDON, F.R.S.E., F.G.S. Demy 8vo. 7*s.* 6*d.*

'The whole subject is dealt with very copiously and clearly in all its parts, and can scarcely fail of appreciation at the hands of practical men, for whose use it is designed.'—*Morning Post.*

'A most valuable treatise.'—*Leeds Mercury.*

'A most valuable addition to a branch of scientific literature, which is of the highest importance not only in a technical aspect but also in a purely scientific view.'
Westminster Review.

The EXPANSE of HEAVEN: a Series of Essays on the
Wonders of the Firmament. By R. A. PROCTOR, B.A. With a Frontispiece. Second Edition. Crown 8vo. 6*s.*

'A very charming work; cannot fail to lift the reader's mind up "through nature's work to nature's God."'—*Standard.*

'Full of thought, readable, and popular.'—*Brighton Gazette.*

OUR PLACE among INFINITIES: with Essays on
'Astrology' and the 'Sabbath of the Jews.' By R. A. PROCTOR, B.A., Author of 'The Expanse of Heaven.' Crown 8vo. 6*s.*

A CLASS BOOK of CHEMISTRY, on the Basis of the
New System. By EDWARD L. YOUMANS, M.D. With 200 Illustrations.

FIRST BOOK of ZOOLOGY. By EDWARD S. MORSE,
Ph.D., late Professor of Comparative Anatomy and Zoology in Bowdoin College. With numerous Illustrations.

An ESSAY on the CULTURE of the OBSERVING POWERS of CHILDREN, especially in Connection with the Study of Botany. By ELIZA A. YOUMANS. Edited, with Notes and a Supplement, by JOSEPH PAYNE, F.C.P., Author of 'Lectures on the Science and Art of Education' &c. Cr. 8vo. 2s.6d.

'This study, according to her just notions on the subject, is to be fundamentally based on the exercise of the pupil's own powers of observation. He is to see and examine the properties of plants and flowers at first hand, not merely to be informed of what others have seen and examined.'—*Pall Mall Gazette.*

FIRST BOOK of BOTANY. Designed to cultivate the Observing Powers of Children. By ELIZA A. YOUMANS. With 300 Engravings. New and Enlarged Edition. Crown 8vo. 5s.

'It is but rarely that a school-book appears which is at once so novel in plan, so successful in execution, and so suited to the general want, as to command universal and unqualified approbation, but such has been the case with Miss Youmans' First Book on Botany. . . . It has been everywhere welcomed as a timely and invaluable contribution to the improvement of primary education.'—*Pall Mall Gazette.*

OBSERVATIONS of MAGNETIC DECLINATION made at TREVANDRUM and AGUSTIA MALLEY in the Observatories of his Highness the Maharajah of Travancore, G.C.S.I., in the years 1852 to 1860. Being Trevandrum Magnetical Observations, Volume I. Discussed and Edited by JOHN ALLEN BROUN, F.R.S., late Director of the Observatories. With an Appendix. Imp. 4to. cloth, £3. 3s.

*** The Appendix, containing Reports on the Observatories and on the Public Museum, Public Park and Gardens at Trevandrum, pp. xii. 116, may be had separately. 21s.

'The title of the work, which is a handsome volume, quarto, 600 pages, at first sight would appear to indicate a dry collection of tables and figures. Some of these, of course, are necessary ; but in addition to them, there is a considerable amount of most interesting matter to the general reader in the descriptions of the adventures and troubles of a scientific man in Southern India, while the magnetician and physicist will find much to occupy his attention in the various results which Mr. Broun has so clearly brought out in his discussion of the observations, and in the description of the very ingenious instruments he constructed and employed in his researches.'
G. M. WHIPPLE, in the *Academy.*

CURRENCY and BANKING. By Professor BONAMY PRICE, Professor of Political Economy in the University at Oxford. 1 vol. crown 8vo. 6s.

The HISTORY of the EVOLUTION of MAN. By Professor ERNST HAECKEL. Translated by E. A. VAN RHYN and L. ELSBERG, M.D. University of New York, with Notes and Additions sanctioned by the Author. Post 8vo.

The HISTORY of CREATION: a Popular Account of the Development of the Earth and its Inhabitants, according to the Theories of Kant, Laplace, Lamarck, and Darwin. By Professor ERNST HAECKEL, of the University of Jena. The Translation revised by E. RAY LANKESTER, M.A., F.R.S. With Coloured Plates and Genealogical Trees of the various Groups of both Plants and Animals. 2 vols. post 8vo. 32s.

HANDBOOK of COMPARATIVE ANATOMY. By Professor OSCAR SCHMIDT, Strasburg University. Crown 8vo.

MANKIND: a Scientific Study of the Races and Distribution of Man, considered in their Bodily Variations, Languages, Occupations, and Religions. By Dr. PESCHEL.

A TREATISE on RELAPSING FEVER. By R. T. LYONS, Assistant-Surgeon, Bengal Army. Post 8vo. 7s. 6d.

'A practical work, thoroughly supported in its views by a series of remarkable cases.'—*Standard.*

CONTEMPORARY ENGLISH PSYCHOLOGY. By Professor TH. RIBOT. Large post 8vo. 9s. An analysis of the views and opinions of the following metaphysicians, as expressed in their writings :—James Mill, Alexander Bain, John Stuart Mill, George H. Lewes, Herbert Spencer, Samuel Bailey.

'The task which M. Ribot set himself he has performed with very great success.'
Examiner.

'We can cordially recommend the volume.'—*Journal of Mental Science.*

HEREDITY : a Psychological Study of its Phenomena, its Laws, its Causes, and its Consequences. By Professor TH. RIBOT. 1 vol. Large crown 8vo. 9s.

HEALTH and DISEASE, as influenced by the Daily, Seasonal, and other Cyclical Changes in the Human System. By EDWARD SMITH, M.D., LL.B., F.R.S. A New Edition. 7s. 6d.

PRACTICAL DIETARY for FAMILIES, SCHOOLS, and the LABOURING CLASSES. By EDWARD SMITH, M.D., LL.B., F.R.S. A New Edition. 3s. 6d.

CONSUMPTION in its EARLY and REMEDIABLE STAGES. By EDWARD SMITH, M.D., LL.B., F.R.S. A New Edition. 7s. 6d.

SENSATION and INTUITION. By JAMES SULLY. Demy 8vo. 10s. 6d.

'The materials furnished by a quick and lively natural sense are happily ordered by a mind trained in scientific method. This merit is especially conspicuous in those parts of the book where, with abundant ingenuity and no mean success, Mr. Sully endeavours to throw some light of cosmic order into the chaos of æsthetics.'
Saturday Review.

'The writer of such an essay (on Belief) must be ranked as a psychologist of no common order.'—Professor BAIN in the *Fortnightly Review.*

'His remarkable collection of studies in psychology and æsthetics. . . . . . Two essays concerned with the æsthetic aspects of human character and its artistic representation display a fine critical tact joined to no common analytical power.'
Professor CROOM ROBINSON in the *Examiner.*

The **PLACE of the PHYSICIAN.** Being the Introductory Lecture at Guy's Hospital, 1873-74 ; to which is added ESSAYS ON THE LAW OF HUMAN LIFE, AND ON THE RELATION BETWEEN ORGANIC AND INORGANIC WORLDS. By JAMES HINTON, late Aural Surgeon to Guy's Hospital. Crown 8vo. 3s. 6d.

PHYSIOLOGY for PRACTICAL USE. By various Writers. Edited by JAMES HINTON. Second Edition. With 50 Illustrations. 2 vols. crown 8vo. 12s. 6d.

'We never saw the popular side of the science of physiology better explained than it is in these two thin volumes.'—*Standard.*

'We heartily commend these essays as setting forth in a style that may be comprehended by all, a number of physiological lessons that ought to be learned by the public.'—*Medical Press.*

'This is an admirable collection of essays on various physiological subjects, written for the most part in such a manner as to be perfectly intelligible to the general reader.'
Land and Water.

'We strongly recommend this book to non-professional readers, from the lucid and logical manner in which the physiological problems of everyday life are stated.'
Nature.

The **QUESTIONS of AURAL SURGERY.** With Illustrations. By JAMES HINTON. Post 8vo. 12s. 6d.

'"The Questions of Aural Surgery" more than maintains the author's reputation as a careful clinician, a deep and accurate thinker, and a forcible and talented writer.'
Lancet.

An **ATLAS** of **DISEASES** of the **MEMBRANA** TYMPANI. With Descriptive Text. By JAMES HINTON. Post 8vo. £6. 6s. The Drawings in this volume are entirely done by hand, and the author is able to say of them that they have been executed with a fidelity and perfectness that has at least equalled his most sanguine expectations.

'By far the best and most accurate that has ever yet been published. The drawings are taken from actual specimens, and are all coloured by hand.'—*Lancet.*

CHOLERA : how to **Avoid** and **Treat** it. Popular and Practical Notes. By HENRY BLANC, M.D. Crown 8vo. 4s. 6d.

'A very practical manual, based on experience and careful observation, full of excellent hints on a most dangerous disease.'—*Standard.*

The **PRINCIPLES** of **MENTAL PHYSIOLOGY.** With their Applications to the Training and Discipline of the Mind, and the Study of its Morbid Conditions. By W. B. CARPENTER, LL.D., M.D., F.R.S., &c. Illustrated. 8vo. 12s.

'. . . . . We have not dealt with the two main views elaborated in this valuable book, from the first of which, together with the inferences which Dr. Carpenter draws as to the sources of our knowledge of necessary truth, we mainly dissent, but with the latter of which we cordially agree. Let us add that nothing we have said, or in any limited space could say, would give an adequate conception of the valuable and curious collection of facts bearing on morbid mental conditions, the learned physiological exposition, and the treasure-house of useful hints for mental training which make this large and yet very amusing, as well as instructive book, an encyclopædia of well-classified and often very startling psychological experiences.'—*Spectator.*

The **COMMON-SENSE MANAGEMENT** of the STOMACH. By G. OVEREND DREWRY, M.D. Fcap. 8vo. Price 2s. 6d.

'This little book appears to us to be a very excellent manual on a highly important subject. It is written in a clear and simple style, and in its occasional touches of humour recals the manner of the famous Abernethy.'—*Graphic.*

'It is well written, sensible and practical rather than scientific, and may be read with profit by every one whether in or out of health. If we are well, he tells us, if we don't know it already, how to keep well.'—*Glasgow Herald.*

'A trustworthy family guide for the preservation of health.'—*Nonconformist.*

'Dr. Drewry has written a little book on the management of the stomach, which certainly justifies the promise of common sense conveyed in the title.'
Saturday Review.

'The author holds common sense views on the subject on which he has written ; and as he has stated many important facts in very plain language, we can recommend his book as a sound and safe guide to the dyspeptic.'—*Morning Advertiser.*

'The principles in self-management according to his common sense manual are undoubtedly sound.'—*Record.*

LONGEVITY; the Means of Prolonging Life after

Middle Age. By JOHN GARDNER, M.D. Third Edition, revised and enlarged. Small crown 8vo. 4*s*.

' Much useful information will be found in its pages.'—*Medical Press.*

' The hints here given on health, and suggestions on longevity, evidently based on experience of a lengthened extent, are, to our mind, invaluable.'—*Standard.*

' Dr. Gardner's directions are sensible enough, and founded on good principles. The advice given is such that any man in moderate health might follow it to advantage, whilst no clap-trap is introduced which might savour of quackery.'

Lancet.

' It is full of valuable information, and abounds in a very uncommon qualit common sense.'—*Edinburgh Daily Review.*